REAL SCIENCE-4-KIDS

KOGS-4-KIDS

BIOLOGY
CONNECTS
TO LANGUAGE

I0030783

Rebecca W. Keller, Ph.D.

Cover design: Rebecca W. Keller, Ph.D.
Opening page: Rebecca W. Keller, Ph.D.
Illustrations: Rebecca W. Keller, Ph.D.

Copyright © 2006, 2011, 2013 Gravitas Publications, Inc.

All rights reserved. No part of this publication may be reproduced, stored in a retrieval system, or transmitted, in any form or by any means, electronic, mechanical, photocopying, recording, or otherwise, without prior written permission from the publisher. No part of this book may be used or reproduced in any manner whatsoever without written permission. However, this publication may be photocopied without permission from the publisher if the copies are to be used only for teaching purposes within a family.

Biology Connects To Language
ISBN # 978-1-936114-98-6

Published by Gravitas Publications, Inc.
www.gravitaspublications.com

Printed in United States

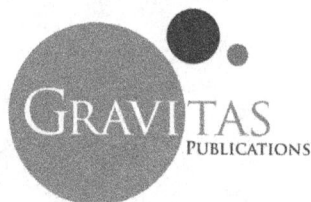

GRAVITAS
PUBLICATIONS

Contents

Introduction

I.1 The language of science

Have you ever noticed that scientists use all kinds of fancy words like nucleosynthesis (nü-klē-ō-sin'-thə-səs) or photoelectric photometry (fō-tō-i-lek'-trik fō-tä'-mə-trē)? Many of the words that scientists use are long and difficult to pronounce. However, these words have been carefully selected by scientists as they put the field of science into verbal language. Each scientific word means a particular thing. There is a *language* to science.

I.2 Latin and Greek roots

If someone had to memorize all of the words that scientists use, it would be a difficult task. Most of the words that are used come to us from two languages: Latin and Greek. Many of the words you will encounter in science will have Latin or Greek *word roots*. A word root is that part of a word that is derived from another word. For example, the word "biology" comes from the Greek word *bios*, which means "life," and *logy* which means "study of," so biology means "the study of life." The word tree illustration

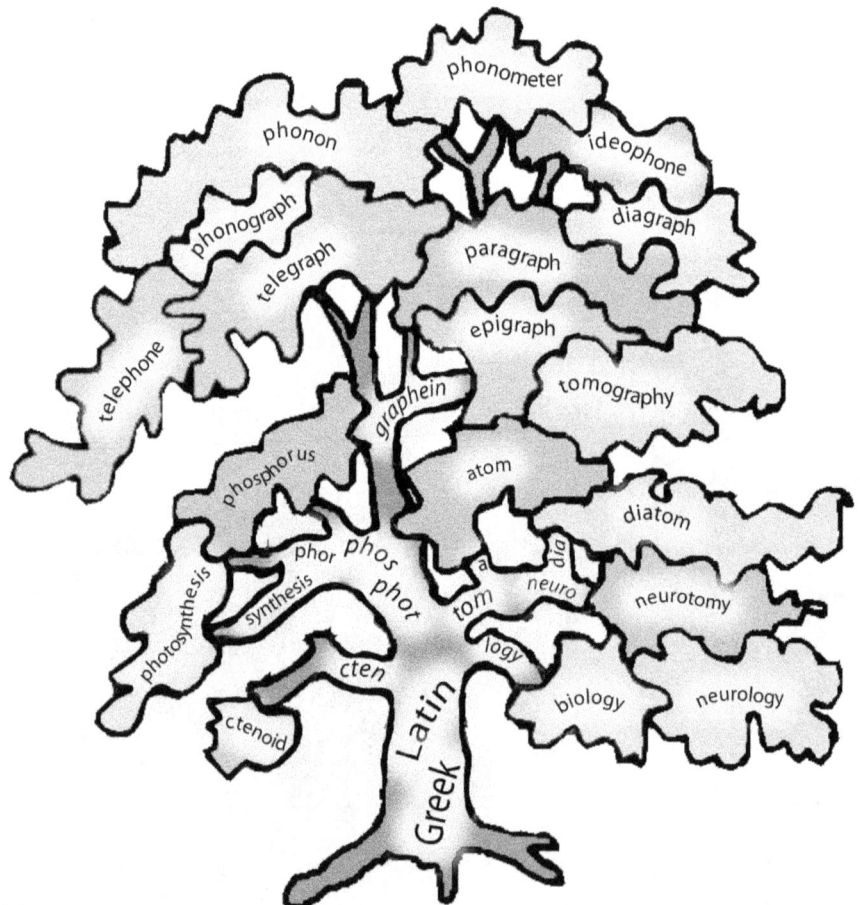

phonometer
phonon
ideophone
phonograph
diagraph
telegraph
paragraph
epigraph
telephone
tomography
graphein
phosphorus
atom
diatom
phor phos
a dia
neuro
neurotomy
photosynthesis
synthesis
phot
tom
logy
cten
biology
neurology
ctenoid
Latin
Greek

shows several different words and their Latin or Greek word roots. You can see that the words on the branches are the word roots, and those on the leaves come from these roots. In fact, many of the languages that people speak have Latin or Greek word roots. English, Spanish, Portuguese, German, and even Romanian all have some words that are similar because some words in each of these languages come from Latin or Greek. For example, the English, Spanish, Portuguese, French, and Italian words for school all come from the Latin word *schola*.

"school"	Language
schola	**Latin**
escuela	Spanish
escola	Portuguese
scuola	Italian
school	English
ecole	French

We can learn a lot about languages by learning the Latin and Greek word roots.

I.3 How this book works

This workbook is called a "connection" because it *connects* the discipline of science to the language of science. Using this workbook will help you to more easily understand the different science subjects because you will know more about the language of science.

In *Find the Root*, the first section of each chapter, you will find the word root in a set of six English words. The pronunciation of each word is shown, and there is a pronunciation key at the end of the book.

For example, your word list may look something like this:

> **deflate** (di-flāt')
>
> **inflate** (in-flāt')
>
> **flabellum** (flə-bel'-əm)
>
> **flavor** (flā'-vər)
>
> **conflate** (kən-flāt')
>
> **afflate** (ə-flāt')

The word root can be three letters long, four letters long, or even five letters long. You are asked to look for the three to five letters (the cluster) that are common in each word. This is the **word root**. For example, this list of words has the following three letters in common:

> **deflate**
>
> **inflate**
>
> **flabellum**
>
> **flavor**
>
> **conflate**
>
> **afflate**

We see that all of the words have the common cluster of letters "f," "l," and "a" which make up the word root "**fla**."

The exercises in the first section of each chapter are designed to get you thinking about the words in the list and their common word root. You are asked to *guess* the meaning of the word root and any of the words in the list.

In *Learn the Root*, the second section of each chapter, the meaning of the word root is defined. For example, we found that the word root for this list is the three letter word root **fla**.

de**fla**te   **fla**vor

in**fla**te   con**fla**te

flabellum   af**fla**te

In this example we discover that the meaning of the word root **fla** comes from the Latin word *flare* which means "wind" or "to blow." All of the words in the list have something to do with "wind" or "blowing." In this section you will be encouraged to try to define the words in the list *before* you look at the definitions. Guessing is good! It gets you thinking, even if your guesses are wrong.

The third section, *Definitions*, defines all the words in the list. These definitions are taken from a number of different dictionaries including *Webster's Unabridged New Twentieth Century Dictionary 1972*, the *American Dictionary of the English Language 1828*, and a *Thesaurus of Word Roots of the English Language*. Additional Latin or Greek word roots for prefixes or suffixes are also given.

The meanings of the words in our example are:

deflate
1. to collapse by letting out air or gas
2. to lessen the importance of, as with money
(*de*—opposite)

inflate
1. to blow full of air, to expand
2. to raise the spirits of
3. to increase, raise beyond normal
(*in*—in)

flabellum   1. a large fan, usually carried by the Pope
                       2. in zoology or botany, a fan-shaped part or
                           structure

flavor   an odor, smell, or aroma *carried by the wind*

conflate   1. blow together, bring together, collect
                       2. to combine, melt, fuse, or join
                       (*con*—together)

afflate   to blow or breathe upon
                       (*af*—to, toward)

In the next section, *Mix and Match*, you are given the opportunity to match the definitions of the words to the word list. For example:

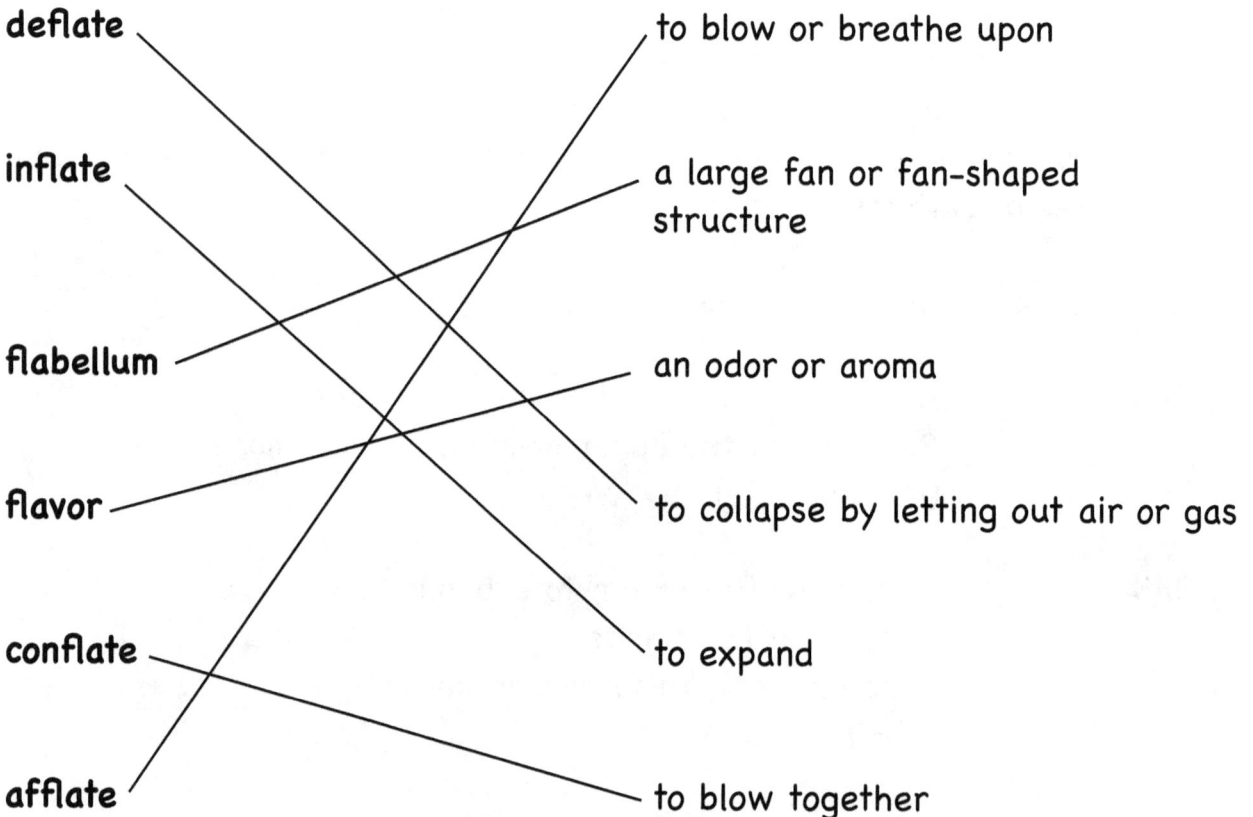

deflate   to blow or breathe upon

inflate   a large fan or fan-shaped
                                         structure

flabellum   an odor or aroma

flavor   to collapse by letting out air or gas

conflate   to expand

afflate   to blow together

Now that you have learned the word root and the definitions of the different words in the list, you can use the *Test Yourself* section to see how well you remember them.

Finally, in the last section of each chapter, *Using New Words,* you will have a chance to make up a story or several sentences using the words you have learned.

For example:

*While carrying a rather **deflated** balloon down the sidewalk, and his hand **conflated** with his mother's, little Johnny was losing his interest with his mother's conversation with Mr. Longs, who awkwardly carrying his **flabellum**, was nevertheless chatting whole-heartedly. The **flavor** of his chocolate sundae was wearing off and his patience was at an end. **Afflating** hard upon his mother, which was his favorite way of gaining her attention, his mother made her farewells and Mr. Longs, still awkwardly carrying his **flabellum**, moved on. Johnny was saved from the boring conversation at last.*

written by Christopher Keller, age 9

Now that you have learned about how language connects to science, use this workbook as you learn Biology Level I, and most importantly...

Have fun!

Chapter 1: Biology

1.1 Find the root

Look at the following words:

biology (bī-ä′-lə-jē)

abiogenesis (ā-bī-ō-je′-nə-səs)

amphibious (am-fi′-bē-əs)

biochemistry (bī-ō-ke′-mə-strē)

autobiography (ô-tə-bī-ä′-grə-fē)

symbiosis (sim-bē-ō′-səs)

There is a cluster, or group, of letters that is exactly the same in all six words. Can you find the cluster?

Circle the cluster that is the same in each word. Write the three letters that make up the cluster. _____ _____ _____

Now look at the words carefully. Because they all have a common cluster, they all have meanings with some similarities.

Can you guess the meaning of the cluster?

_____

_____

Can you write a definition for any of the words in the list?

_____

_____

_____

1.2 Learn the root

Word Cluster

biology   a**bio**genesis

amphi**bio**us   **bio**chemistry

auto**bio**graphy   sym**bio**sis

All of the words above have a common word root—**bio**. The word root **bio** comes from the Greek word *bios*, which means "to live" or"life." All of the words have something to do with "living" or "life."

Now, knowing that the cluster **bio** comes from the Greek word *bios*, try to guess the meanings of the words before looking at the definitions in the next section.

biology _____

abiogenesis _____

amphibious _____

biochemistry _____

autobiography _____

symbiosis _____

1.3 Definitions

biology	The field of science concerned with living things. (*logos*—to gather; to study)
abiogenesis	literally means "life without production." Abiogenesis is also known as spontaneous generation and is the idea that living organisms can come directly from lifeless matter. This theory has been rejected as a result of the pioneering work by Louis Pasteur who, in the late 1800s, showed that "life comes from life." (*a*—without, or not; *genus*—to produce)
amphibious	to live a double life; in biology, creatures that live both on land and in the water are amphibious (*amphi*—on both sides)
biochemistry	The branch of chemistry that deals primarily with the chemistry of living things.
autobiography	The life story of a person written by that person. (*auto*—self; *graphy*—to write)
symbiosis	the living together of different creatures under the same condition or in close association (*sym*—together; *osis*—condition).

1.4 Mix and match

Draw lines to connect the words with their meanings.

biology

the living together of different creatures under the same condition or in close association

abiogenesis

field of science concerned with living things

amphibious

spontaneous generation

biochemistry

the life story of a person written by that person

autobiography

living a double life

symbiosis

a branch of chemistry dealing with biological systems

1.5 Test yourself

Write the meaning next to each word below.

biology _____

abiogenesis _____

amphibious _____

biochemistry _____

autobiography _____

symbiosis _____

Extra:

Can you guess the meanings of the following words?

bioethics (*ethos*—moral custom) (bī-ō-e′-thiks)

_____

bioherm (*herma*—reef) (bī′-ō-herm)

_____

1.6 Using new words

Write several sentences using each new word you have learned.

_____

_____

_____

_____

_____

_____

_____

_____

_____

_____

_____

_____

Chapter 2: Chloroplast

2.1 Find the root

Look at the following words:

> **chloroplast** (klôr'-ə-plast)
>
> **plastic** (plas'-tik)
>
> **protoplast** (prō'-tə-plast)
>
> **neoplastic** (nē-ə-plas'-tik)
>
> **plaster** (plas'-tər)
>
> **chromoplast** (krō'-mə-plast)

There is a cluster of letters that is exactly the same in all six words. Can you find the cluster?

Circle the cluster that is the same in each word. Write the five letters that make up the cluster. _____ _____ _____ _____ _____

Now look at the words carefully. Because they all have a common cluster, they all have meanings with some similarities.

Can you guess the meaning of the cluster?

_____

_____

Can you write a definition for any of the words in the list?

_____

_____

_____

2.2 Learn the root

Word Cluster

chloro**plast**   **plast**ic

proto**plast**   neo**plast**ic

plaster   chromo**plast**

All of the words above have a common word root—**plast**. The word root **plast** comes from the Greek word *plassein*, which means "to mold or form." All of the words have something to do with "molding" or "forming."

Now, knowing that the cluster **plast** comes from the Greek word *plassein*, try to guess the meanings of the words before looking at the definitions in the next section.

chloroplast _____

plastic _____

protoplast _____

neoplastic _____

plaster _____

chromoplast _____

2.3 Definitions

chloroplast   the organelle inside plant cells that converts [molds, shapes, forms] light energy into food (*chloro*—green)

plastic   *noun:* a man-made, non-metallic compound that can be molded into various forms
adjective: capable of being molded or shaped

protoplast   a thing that is the first of its kind; the first "mold"

neoplastic   having to do with any new or abnormal growth or tissue; relating to a tumor

plaster   a mixture of sand, lime and water used for coating walls and ceilings; *Plaster of Paris*—a white powder used to make molds for casting, or to make sculptural shapes

chromoplast   in biology, any colored body or shape found in a cell

2.4 Mix and match

Draw lines to connect the words with their meanings.

chloroplast   the first of its kind

plastic   relating to a tumor

protoplast   something that can be molded or
                                   shaped

neoplastic   an organelle inside plant cells that
                                   converts light energy to food

plaster   a colored body or shape found inside
                                   cells

chromoplast   a material made of sand and water
                                   used for coating ceilings and walls

2.5 Test yourself

Write the meaning next to each word below.

chloroplast _____

plastic _____

protoplast _____

neoplastic _____

plaster _____

chromoplast _____

Extra:

Can you guess the meanings of the following words?

dermatoplasty (*derm*—skin) (der-mat´-ə-plas-tē)

_____

thermoplastic (*therm*—heat) (thər-mə-plas´-tik)

_____

2.6 Using new words

Write several sentences using each new word you have learned.

_____

_____

_____

_____

_____

_____

_____

_____

_____

_____

_____

Chapter 3: Photosynthesis

3.1 Find the root

Look at the following words:

photosynthesis (fō-tō-sin′-thə-səs)

syndicate (sin′-də-kət)

synonym (si′-nə-nim)

biosynthesis (bī-ō-sin′-thə-səs)

idiosyncrasies (i-dē-ə-sin′-krə-sēs)

asyndeton (ə-sin′-də-tän)

There is a cluster of letters that is exactly the same in all six words. Can you find the cluster?

Circle the cluster that is the same in each word. Write the three letters that make up the cluster. _____ _____ _____

Now look at the words carefully. Because they all have a common cluster, they all have meanings with some similarities.

Can you guess the meaning of the cluster?

_____

_____

Can you write a definition for any of the words in the list?

_____

_____

_____

3.2 Learn the root

Word Cluster

photo**syn**thesis   **syn**dicate

synonym   bio**syn**thesis

idio**syn**crasies   a**syn**deton

All of the words above have a common word root—**syn**. The word root **syn** comes from the Greek word *syndein*, which means "with, together with, at the same time, or same." All of the words have something to do with "together" or "same."

Now, knowing that the cluster **syn** comes from the Greek word *syndein*, try to guess the meanings of the words before looking at the definitions in the next section.

photosynthesis _____

syndicate _____

synonym _____

biosynthesis _____

idiosyncrasies _____

asyndeton _____

3.3 Definitions

photosynthesis   literally, "making together with light"; the process by which plants convert light energy into food (*photo*—light; *thesis*—to make)

syndicate   a group that comes together to work toward achieving a goal

synonym   one of two or more words having the same or nearly the same meaning (*nomen*—name)

biosynthesis   literally—"making together with life"; any process by which living things produce a product (*bio*—life)

idiosyncrasies   one's own individual peculiarities or mannerisms that are found together as a group

asyndeton   the practice of leaving out conjunctions between sentence elements; literally "not together with" (*a*—without); e.g., the Latin sentence, *Veni, vidi, vici*—"I came, I saw, I conquered."

3.4 Mix and match

Draw lines to connect the words with their meanings.

photosynthesis   a group that comes together to
                                            work toward achieving a goal

syndicate   any process by which living things
                                            produce a product

synonym   one's own peculiarities found
                                            together as a group

biosynthesis   the process that plants use to
                                            convert light energy to food

idiosyncrasies   the leaving out of conjunctions

asyndeton   one of two or more words having
                                            the same meaning

3.5 Test yourself

Write the meaning next to each word below.

photosynthesis _____

syndicate _____

synonym _____

biosynthesis _____

idiosyncrasies _____

asyndeton _____

Extra:

Can you guess the meanings of the following words?

synagogue (*agog*—leading) (si'-nə-gäg)

_____

geosynchronous (*chronos*—time) (jē-ō-sin'-krə-nəs)

_____

3.6 Using new words

Write several sentences using each new word you have learned.

_____

_____

_____

_____

_____

_____

_____

_____

_____

_____

_____

_____

Chapter 4: Carpel

4.1 Find the root

Look at the following words:

carpel (kär'-pəl)

epicarp (e'-pi-kärp)

carpology (kärp-ä'-lə-jē)

carpophore (kär'-pə-fôr)

schizocarp (ski'-zə-kärp)

monocarp (mä'-nə-kärp)

There is a cluster of letters that is exactly the same in all six words. Can you find the cluster?

Circle the cluster that is the same in each word. Write the four letters that make up the cluster. _____ _____ _____ _____

Now look at the words carefully. Because they all have a common cluster, they all have meanings with some similarities.

Can you guess the meaning of the cluster?

_____

_____

Can you write a definition for any of the words in the list?

_____

_____

_____

4.2 Learn the root

Word Cluster

carpel   epi**carp**

carpology   **carp**ophore

schizo**carp**   mono**carp**

All of the words above have a common word root—**carp**. The word root **carp** comes from the Greek word *karpos*, which means "fruit." All of the words have something to do with "fruit."

Now, knowing that the cluster **carp** comes from the Greek word *karpos*, try to guess the meanings of the words before looking at the definitions in the next section.

carpel _____

epicarp _____

carpology _____

carpophore _____

schizocarp _____

monocarp _____

4.3 Definitions

carpel

diminutive of *karpos*, meaning "little fruit"; the part of a flower where seeds are made by the plant; regarded as a modified leaf

epicarp

the outer layer (skin) of a ripened fruit
(*epi*—upon)

carpology

the study of fruits and seeds
(*logy*—study of)

carpophore

in botany, the organ that supports the carpels
(*pherin*—to bear)

schizocarp

a dry fruit that splits into two or more one-seeded carpels at maturity
(*schizo*—split)

monocarp

a plant that bears fruit only once and then dies; e.g., annuals, biennials, bamboo
(*mono*—one)

4.4 Mix and match

Draw lines to connect the words with their meanings.

carpel   the study of fruits and seeds

epicarp   a dry fruit that splits into two or more
                                    one-seeded carpels

carpology   the skin of a ripened fruit

carpophore   a plant that bears fruit only once and
                                    then dies

schizocarp   "little fruit"; the part of a plant where
                                    seeds are made

monocarp   the organ that supports the carpel

4.5 Test yourself

Write the meaning next to each word below.

carpel _____

epicarp _____

carpology _____

carpophore _____

schizocarp _____

monocarp _____

Extra:

Can you guess the meanings of the following words?

polycarpic (*poly*—many) (pä-lē-kärp'-ik)

_____

carpophagous (*phagein*—to eat) (kär-pä'-fə-gəs)

_____

4.6 Using new words

Write several sentences using each new word you have learned.

_____

_____

_____

_____

_____

_____

_____

_____

_____

_____

_____

Chapter 5: Germination

5.1 Find the root

Look at the following words:

germination (jər-mə-nā′-shən)

monogerm (mä′-nə-jərm)

german (jər′-mən)

multigerm (məl-tē-jərm′)

ungerminated (un-jər′-mə-nā-təd)

germicide (jər′-mə-sīd)

There is a cluster of letters that is exactly the same in all six words. Can you find the cluster?

Circle the cluster that is the same in each word. Write the four letters that make up the cluster. _____ _____ _____ _____

Now look at the words carefully. Because they all have a common cluster, they all have meanings with some similarities.

Can you guess the meaning of the cluster?

_____

_____

Can you write a definition for any of the words in the list?

_____

_____

_____

5.2 Learn the root

Word Cluster

germination   mono**germ**

german   multi**germ**

un**germ**inated   **germ**icide

All of the words above have a common word root—**germ**. The word root **germ** comes from the Latin word *gignere*, which means "to beget" or "to sprout" or "to bud." All of the words have something to do with "begetting" or "sprouting" or "budding."

Now, knowing that the cluster **germ** comes from the Latin word *gignere*, try to guess the meanings of the words before looking at the definitions in the next section.

germination _____

monogerm _____

german _____

multigerm _____

ungerminated _____

germicide _____

5.3 Definitions

germination   to cause to develop, sprout, or come into being

monogerm   a fruit that gives rise to a single plant
(*mono*—one)

german   having the same parents or having the same
grandparents on either the maternal or paternal side

multigerm   a fruit that gives rise to several plants
(*multi*—many)

ungerminated   a seed that has not yet sprouted
(*un*—not)

germicide   something that kills "germs"
(*cide*—to kill)

5.4 Mix and match

Draw lines to connect the words with their meanings.

germination   something that kills "germs"

monogerm   having the same parents or
                                   grandparents

german   a seed that has not yet sprouted

multigerm   a fruit that gives rise to a single
                                   plant

ungerminated   to cause to develop, sprout, or come
                                   into being

germicide   a fruit that gives rise to several plants

5.5 Test yourself

Write the meaning next to each word below.

germination _____

monogerm _____

german _____

multigerm _____

ungerminated _____

germicide _____

Extra:

Can you guess the meanings of the following words?

germiculture (*cultus*—to grow, to care for, cultivate) (jərm′-ə-kəl-chər)

_____

germinal (*-al*—to be like) (jər′-mə-nəl)

_____

5.6 Using new words

Write several sentences using each new word you have learned.

_____

_____

_____

_____

_____

_____

_____

_____

_____

_____

_____

_____

_____

Chapter 6: Pseudopodia

6.1 Find the root

Look at the following words:

pseudopodia (sü-də-pō´-dē-ə)

podium (pō´-dē-əm)

podiatry (pə-dī´-ə-trē)

arthropod (är´-thrə-päd)

tripod (trī´-päd)

octopod (äk´-tə-päd)

There is a cluster of letters that is exactly the same in all six words. Can you find the cluster?

Circle the cluster that is the same in each word. Write the three letters that make up the cluster. _____ _____ _____

Now look at the words carefully. Because they all have a common cluster, they all have meanings with some similarities.

Can you guess the meaning of the cluster?

_____

_____

Can you write a definition for any of the words in the list?

_____

_____

_____

6.2 Learn the root

Word Cluster

pseudo**pod**ia   **pod**ium

podiatry   arthro**pod**

tri**pod**   octo**pod**

All of the words above have a common word root—**pod**. The word root **pod** comes from the Greek word *pous*, which means "foot." All of the words have something to do with "foot."

Now, knowing that the cluster **pod** comes from the Greek word *pous*, try to guess the meanings of the words before looking at the definitions in the next section.

pseudopodia _____

podium_____

podiatry _____

arthropod _____

tripod _____

octopod_____

6.3 Definitions

pseudopodia in zoology, a temporary projection from a single-celled organism that is used as a foot
(*pseudo*—false)

podium 1. a low wall serving as a pedestal or foundation
2. a raised platform for a conductor or speaker

podiatry the branch of medicine having to do with the care and treatment of feet and their disorders

arthropod a member of the phylum Arthopoda which includes creatures such as spiders and lobsters that have segmented bodies and "feet"
(*arthro*—jointed)

tripod 1. a stool, table, or seat with three legs.
2. a frame or support with three legs
(*tri*—three)

octopod a sea creature that has eight legs (feet) and is in the phylum Mollusca and the order Octopoda; includes the octopus
(*octo* — eight)

6.4 Mix and match

Draw lines to connect the words with their meanings.

pseudopodia   creatures with jointed feet, such as
                                   spiders or lobsters

podium   sea creature with eight legs

podiatry   a stool or table with three legs

arthropod   the branch of medicine that deals with
                                   feet and their problems

tripod   in single-celled organisms, a false foot

octopod   a raised platform for a conductor or
                                   speaker

6.5 Test yourself

Write the meaning next to each word below.

pseudopodia _____

podium _____

podiatry _____

arthropod _____

tripod _____

octopod _____

Extra:

Can you guess the meanings of the following words?

apodal (*a*—without) (ap'-əd-əl)

_____

bipod (*bi*—two) (bī'-päd)

_____

6.6 Using new words

Write several sentences using each new word you have learned.

_____

_____

_____

_____

_____

_____

_____

_____

_____

_____

_____

Chapter 7: Phagocytosis

7.1 Find the root

Look at the following words:

phagocytosis (fa-gə-sə-tō'-səs)

aphagia (ə-fā'-jē-ə)

aerophagia (ar-ō-fā'-jē-ə)

anthropophagite (an-thrə-pə-fā'-gīt)

bacteriophage (bak-tir'-ē-ə-fāj)

monophagous (mə-nof'-ə-gəs)

There is a cluster of letters that is exactly the same in all six words. Can you find the cluster?

Circle the cluster that is the same in each word. Write the four letters that make up the cluster. _____ _____ _____ _____

Now look at the words carefully. Because they all have a common cluster, they all have meanings with some similarities.

Can you guess the meaning of the cluster?

_____

_____

Can you write a definition for any of the words in the list?

_____

_____

_____

7.2 Learn the root

Word Cluster

phagocytosis   a**phag**ia

aero**phag**ia   anthropo**phag**ite

bacterio**phage**   mono**phag**ous

All of the words above have a common word root—**phag**. The word root **phag** comes from the Greek word *phagein*, which means "to eat." All of the words have something to do with "eating."

Now, knowing that the cluster **phag** comes from the Greek word *phagein*, try to guess the meanings of the words before looking at the definitions in the next section.

phagocytosis _____

aphagia _____

aerophagia _____

anthropophagite _____

bacteriophage _____

monophagous _____

7.3 Definitions

phagocytosis   the action or process of any cell eating other cells

aphagia   in medicine, the loss of the power to swallow

aerophagia   abnormal, spasmodic swallowing or gulping of air
                          (*aero*—air)

anthropophagite   a man-eater; cannibal
                          (*anthropo*—man).

bacteriophage   a microscopic organism that destroys and eats
                          bacteria.

monophagous   eating only one kind of food
                          (*mono*—one)

7.4 Mix and match

Draw lines to connect the words with their meanings.

phagocytosis   a cannibal

aphagia   abnormal gulping of air

aerophagia   the loss of the power to swallow

anthropophagite   eating only one kind of food

bacteriophage   the action or process of any cell eating
                                    another cell

monophagous   a microscopic organism that eats
                                    bacteria

7.5 Test yourself

Write the meaning next to each word below.

phagocytosis _____

aphagia _____

aerophagia _____

anthropophagite _____

bacteriophage _____

monophagous _____

Extra:

Can you guess the meanings of the following words?

entomophagous (*entom*—insect) (en-tə-mä′-fə-gəs)

_____

xylophagous (*xylo*—wood) (zī-lä′-fə-gəs)

_____

7.6 Using new words

Write several sentences using each new word you have learned.

_____

_____

_____

_____

_____

_____

_____

_____

_____

_____

_____

Chapter 8: Metamorphosis

8.1 Find the root

Look at the following words:

metamorphosis (met-ə-môr´-fə-səs)

amorphous (ə-môr´-fəs)

morphology (môr-fäl´-ə-jē)

morphophone (môr´-fə-fōn)

dimorphism (dī-môr´-fi-zəm)

Morpheus (môr´-fē-əs)

There is a cluster of letters that is exactly the same in all six words. Can you find the cluster?

Circle the cluster that is the same in each word. Write the five letters that make up the cluster. _____ _____ _____ _____ _____

Now look at the words carefully. Because they all have a common cluster, they all have meanings with some similarities.

Can you guess the meaning of the cluster?

_____

_____

Can you write a definition for any of the words in the list?

_____

_____

_____

8.2 Learn the root

Word Cluster

meta**morph**osis   a**morph**ous

morphology   **morph**ophone

di**morph**ism   **Morph**eus

All of the words above have a common word root—**morph**. The word root **morph** comes from the Greek word *morphe*, which means "form" or "shape." All of the words have something to do with "form" or "shape."

Now, knowing that the cluster **morph** comes from the Greek word *morphe*, try to guess the meanings of the words before looking at the definitions in the next section.

metamorphosis _____

amorphous _____

morphology _____

morphophone _____

dimorphism _____

Morpheus _____

8.3 Definitions

metamorphosis   a transformation; to change shape, structure, form, or substance; a complete change in appearance or condition
(*meta*—over, but can mean involving change)

amorphous   without shape or form; shapeless, formless, characterless, unorganized
(*a*—without)

morphology   any scientific study of shape or form, such as in biology or linguistics
(*logy*—scientific study of)

morphophone   an early music synthesizer with a specialized loop deck, an erase head, a record head, and ten playback heads with an adjustable filter for each which creates special timbre effects
(*phone*—sound)

dimorphism   having two forms; in botany, having two different kinds of leaves
(*di*—two)

Morpheus   in Greek mythology, the god of dreams who appears in many shapes

8.4 Mix and match

Draw lines to connect the words with their meanings.

metamorphosis   in Greek mythology, the god of dreams

amorphous   having two forms

morphology   without shape or form

morphophone   the scientific study of shape or forms

dimorphism   to change shape or form

Morpheus   an early music synthesizer

8.5 Test yourself

Write the meaning next to each word below.

metamorphosis _____

amorphous _____

morphology _____

morphophone _____

dimorphism _____

Morpheus _____

Extra:

Can you guess the meanings of the following words?

neomorph (*neo*—new) (nē'-ə-môrf)

_____

geomorphology (*geo*—earth) (jē-ə-môr-fä'-lə-jē)

_____

8.6 Using new words

Write several sentences using each new word you have learned.

_____

_____

_____

_____

_____

_____

_____

_____

_____

_____

_____

Chapter 9: Lepidoptera

9.1 Find the root

Look at the following words:

Lepidoptera (le-pə-däp′-tə-rə)

helicopter (he′-lə-käp-tər)

pterodactyl (ter-ə-dak′-təl)

apteryx (ap′-tə-riks)

peripteral (pə-rip′-tər-əl)

pteric (ter′-ik)

There is a cluster of letters that is exactly the same in all six words. Can you find the cluster?

Circle the cluster that is the same in each word. Write the four letters that make up the cluster. _____ _____ _____ _____

Now look at the words carefully. Because they all have a common cluster, they all have meanings with some similarities.

Can you guess the meaning of the cluster?

_____

_____

Can you write a definition for any of the words in the list?

_____

_____

_____

9.2 Learn the root

Word Cluster

Lepido**pter**a   helico**pter**

pterodactyl   a**pter**yx

peri**pter**al   **pter**ic

All of the words above have a common word root—**pter**. The word root **pter** comes from the Greek word *pteron*, which means "wing" or "feather" or "winglike." All of the words have something to do with "wing" or "feather" or "winglike."

Now, knowing that the cluster **pter** comes from the Greek word *pteron*, try to guess the meanings of the words before looking at the definitions in the next section.

Lepidoptera _____

helicopter _____

pterodactyl _____

apteryx _____

peripteral _____

pteric _____

9.3 Definitions

Lepidoptera   the order of insects that have wings with scales; includes butterflies and moths
(*lepido*—scales)

helicopter   a kind of aircraft moved by a large propeller (spiral wing) mounted above the fuselage
(*helikos*—spiral)

pterodactyl   an extinct flying reptile having wings of skin stretched along the body between the hind limb and forelimb
(*daktylos*—finger, toe)

apteryx   a genus of nearly extinct tailless birds in New Zealand that have long slender bills and underdeveloped wings
(*a*—without)

peripteral   in Greek architecture, a building that is surrounded by a row of columns.
(*peri*—around)

pteric   pertaining to a wing or shoulder

9.4 Mix and match

Draw lines to connect the words with their meanings.

Lepidoptera   pertaining to a wing or shoulder

helicopter   a building that is surrounded by a row
                                         of columns

pterodactyl   the order of insects that have wings
                                         with scales

apteryx   an extinct flying reptile

peripteral   a tailless bird with underdeveloped
                                         wings

pteric   an aircraft that has a propeller (spiral
                                         wing) attached to the fuselage

9.5 Test yourself

Write the meaning next to each word below.

Lepidoptera _____

helicopter _____

pterodactyl _____

apteryx _____

peripteral _____

pteric _____

Extra:

Can you guess the meanings of the following words?

macropterous (*macro*—large) (ma-kräp′-tə-rəs)

_____

orthopterous (*ortho*—straight) (ôr-thäp′-tə-rəs)

_____

9.6 Using new words

Write several sentences using each new word you have learned.

_____

_____

_____

_____

_____

_____

_____

_____

_____

_____

_____

_____

_____

Chapter 10: Ecosystem

10.1 Find the root

Look at the following words:

ecosystem (ē'-kō-sis-təm)

ecosphere (ē'-kō-sfir)

ecology (i-kä'-lə-jē)

ecotone (ē'-kə-tōn)

ecocide (ē'-kə-sīd)

ecophobia (ē-kə-fō'-bē-ə)

There is a cluster of letters that is exactly the same in all six words. Can you find the cluster?

Circle the cluster that is the same in each word. Write the three letters that make up the cluster. _____ _____ _____

Now look at the words carefully. Because they all have a common cluster, they all have meanings with some similarities.

Can you guess the meaning of the cluster?

_____

_____

Can you write a definition for any of the words in the list?

_____

_____

_____

10.2 Learn the root

Word Cluster

ecosystem   **eco**sphere

ecology   **eco**tone

ecocide   **eco**phobia

All of the words above have a common word root—**eco**. The word root **eco** comes from the Greek word *oikos*, which means "house or settlement." All of the words have something to do with "house or settlement."

Now, knowing that the cluster **eco** comes from the Greek word *oikos*, try to guess the meanings of the words before looking at the definitions in the next section.

ecosystem _____

ecosphere _____

ecology _____

ecotone _____

ecocide _____

ecophobia _____

10.3 Definitions

ecosystem   a community of plants, animals, bacteria, and other living things housed together

ecosphere   a small-scale ecosystem housed in a spherical globe

ecology   the study of communities of plants, animals, bacteria, and other living things that are housed together
(*logy*—study)

ecotone   the stretch of land where two different environments meet; a transition between two different types of habitats
(*tone*—stretch)

ecocide   the destruction of the environment
(*Latin, caedere*—to kill)

ecophobia   a fear or dislike of one's home
(*phobia*—fear)

10.4 Mix and match

Draw lines to connect the words with their meanings.

ecosystem   a fear of one's home

ecosphere   the stretch of land between two
                                                 different habitats

ecology   the destruction of the environment

ecotone   a community of living things housed
                                                 together

ecocide   a small ecosystem housed in a sphere

ecophobia   the study of communities of living
                                                 things

10.5 Test yourself

Write the meaning next to each word below.

ecosystem _____

ecosphere _____

ecology _____

ecotone _____

ecocide _____

ecophobia _____

Extra:

Can you guess the meanings of the following words?

ecologist (*-logist*—a specialist) (i-kä′-lə-jist)

_____

economics (*nemein*—to distribute or manage) (e-kə-nä′-miks)

_____

10.6 Using new words

Write several sentences using each new word you have learned.

_____

_____

_____

_____

_____

_____

_____

_____

_____

_____

_____

Pronunciation Key

a	add	ī	ice	s	sea
ā	race	j	joy	sh	sure
ä	palm	k	cool	t	take
â(r)	air	l	love	u	up
b	bat	m	move	ü	sue
ch	check	n	nice	v	vase
d	dog	ng	sing	w	way
e	end	o	odd	y	yarn
ē	tree	ō	open	z	zebra
f	fit	ô	jaw	ə	a in above
g	go	oi	oil		e in sicken
h	hope	oo	pool		i in possible
i	it	p	pit		o in melon
		r	run		u in circus

More REAL SCIENCE-4-KIDS Books
by Rebecca W. Keller, PhD

Building Blocks Series
yearlong study program — each Student Textbook has accompanying Laboratory Notebook, Teacher's Manual, Lesson Plan, Study Notebook, Quizzes, and Graphics Package

Exploring Science Book K (Activity Book)
Exploring Science Book 1
Exploring Science Book 2
Exploring Science Book 3
Exploring Science Book 4
Exploring Science Book 5
Exploring Science Book 6
Exploring Science Book 7
Exploring Science Book 8

Focus On Series
unit study program — each title has a Student Textbook with accompanying Laboratory Notebook, Teacher's Manual, Lesson Plan, Study Notebook, Quizzes, and Graphics Package

Focus On Elementary Chemistry
Focus On Elementary Biology
Focus On Elementary Physics
Focus On Elementary Geology
Focus On Elementary Astronomy

Focus On Middle School Chemistry
Focus On Middle School Biology
Focus On Middle School Physics
Focus On Middle School Geology
Focus On Middle School Astronomy

Focus On High School Chemistry

Super Simple Science Experiments

21 Super Simple Chemistry Experiments
21 Super Simple Biology Experiments
21 Super Simple Physics Experiments
21 Super Simple Geology Experiments
21 Super Simple Astronomy Experiments
101 Super Simple Science Experiments

Note: A few titles may still be in production.

Gravitas Publications Inc.
www.gravitaspublications.com
www.realscience4kids.com

GRAVITAS
PUBLICATIONS

www.ingramcontent.com/pod-product-compliance
Lightning Source LLC
Chambersburg PA
CBHW051230200326
41519CB00025B/7315